SPECTRUM MANAGEMENT FOR THE 21st CENTURY

A Report of the CSIS Commission on Spectrum Management

Commission Cochairs
Robert Galvin
James Schlesinger

Project Director
James A. Lewis

October 2003

About CSIS

For four decades, the Center for Strategic and International Studies (CSIS) has been dedicated to providing world leaders with strategic insights on—and policy solutions to—current and emerging global issues.

CSIS is led by John J. Hamre, former U.S. deputy secretary of defense. It is guided by a board of trustees chaired by former U.S. senator Sam Nunn and consisting of prominent individuals from both the public and private sectors.

The CSIS staff of 190 researchers and support staff focus primarily on three subject areas. First, CSIS addresses the full spectrum of new challenges to national and international security. Second, it maintains resident experts on all of the world's major geographical regions. Third, it is committed to helping to develop new methods of governance for the global age; to this end, CSIS has programs on technology and public policy, international trade and finance, and energy.

Headquartered in Washington, D.C., CSIS is private, bipartisan, and tax-exempt. CSIS does not take specific policy positions; accordingly, all views expressed herein should be understood to be solely those of the author(s).

Library of Congress Cataloging-in-Publication Data
CIP information available on request.

ISBN 0-89206-437-4

The CSIS Press
Center for Strategic and International Studies
1800 K Street, N.W., Washington, D.C. 20006
Tel: (202) 887-0200
Fax: (202) 775-3199
E-mail: books@csis.org
Web site: www.csis.org/

Contents

Executive Summary v

Introduction 1

The Need for Change

A National Resource

New Spectrum Technologies Create Opportunity and Risk

Recommendations for Improved Spectrum Management

Recommendation 1. White House Oversight 11

Special Assistant for Spectrum Management

Policy Coordinating Committee for Spectrum Management

Dispute Resolution

Recommendation 2. Spectrum Advisory Board 15

Recommendation 3. Reinforce International Functions 18

Recommendation 4. Research Support for Spectrum Innovation 22

Recommendation 5. Create a National Spectrum Strategy 26

Balancing Private and Public-Service Spectrum Needs

Protecting Safety-of-life Services

Markets or Commons

Conclusion 34

A Checklist for Federal Spectrum Management Reform

Appendix A. Additional Views 37

Appendix B. Participants List 41

Executive Summary

U.S. spectrum management is outmoded. It has not kept pace with changes in technology and markets. New technologies and new services create immense opportunities for our nation, but an outmoded legal and management structure hobbles efforts to capture the benefits of innovation. These new technologies are not without risk, but they offer a substantial opportunity to satisfy the growing demand for new services that has produced intense competition for scarce radio spectrum.

New technologies and services have created rising demand for spectrum. Spectrum is a finite natural resource—we cannot make more—and under our current rules, demand outstrips supply. However, the same technologies that create this demand can provide a solution, by allowing more efficient use of the spectrum. This would meet existing and potential demand and could be the basis for unprecedented economic growth. Our existing organizational and legal structure, inherited from an earlier technological era, blocks the development and adoption of the new spectrum technologies. To solve the spectrum problem and exploit this technological opportunity, spectrum management must change.

The Center for Strategic and International Studies (CSIS) established a commission to assess spectrum management and consider changes in policies and procedures that would better meet the national interest (members are listed on pages 41–43). The goal was to find practical recommendations to replace the existing structure for decisionmaking with a process oriented toward long-term national objectives. This report grows out of the work of this commission. It provides an overview of the issue and recommendations on four key problems for U.S. spectrum management:

- The absence of long-range plans or a vision for spectrum use to guide policy and provide a greater degree of certainty for investors and clarity for innovators;
- The lack of an effective mechanism for resolving disputes among federal entities over spectrum policy;
- The increasing challenges in international spectrum negotiations; and
- The risks to U.S. security and economic growth from a potential lag in the development and use of new technologies.

A New Spectrum Environment

A failure to take full advantage of wireless technologies will hurt the United States. Increased access to radio spectrum is essential for national security, public safety, and economic growth. Spectrum access enables mobility and connectivity, and demand for spectrum continues to increase in all sectors. For the military, advantage increasingly comes from information superiority, and information superiority depends on access to spectrum. Mobile applications and networked sensors are the core of the information capabilities needed for military dominance in the twenty-first century. Operations in Kosovo required 10 times the bandwidth needed for the 1990 Gulf War, even though forces deployed to Kosovo were much smaller than those sent to the Gulf. Operations in Afghanistan and Iraq required more than 40 times the bandwidth used by the United States in the Gulf War.

Access to radio spectrum also plays a crucial role in public safety and homeland security. September 11, 2001, found responding fire and police departments at times unable to communicate with each other by radio. Expanding access and reliability for first responders and public safety networks is a crucial task. Broadcast television and radio play a critical role in the public dissemination of news and information. Safety of flight, which often depends on low-powered signals, also requires spectrum access that is protected from interference. Signals from global positioning satellites and from instrument landing systems are essential links in the air transportation network.

For the economy, technologies that exploit the radio spectrum provide competitive advantage. Industries that generate hundred of billions of dollars depend on spectrum access, and new industries continue to appear. The evolution of communications is leading to a range of new spectrum-using services that will generate intense consumer demand and increase productivity. The economic benefits of spectrum access are immense and, if the United States can organize itself to take advantage of them, a key source of future economic growth.

Innovative spectrum-based technologies are being developed at a rapid pace, and many are already deployed. Many new technologies are still experimental, but others (such as 802.11 "Wi-Fi" devices) are now mass-market commodities. This wave of innovation began with developments in military radio equipment and cellular telephones. They changed how people use spectrum. Wireless technologies developed for the military emphasize mobility, high volume, and resistance to interference in order to avoid interception and jamming. These attributes are commercially desirable as well.

Technological change has led to new ways to transmit and receive radio signals, to digitize radio transmissions and to exploit differences in time and space. This could allow for much more intensive use of the spectrum, alleviating "shortages," and reducing the need for cumbersome regulatory practices—if research and experience show that they can operate without causing harmful interference to existing services that are valuable and often vital. However, these technologies use spectrum in a very different way than the older technologies for which U.S. spectrum policies and the regulatory structures were designed. They require a different approach to policy and regulation.

The physical characteristics of radio spectrum also intensify competition. Different parts of the spectrum have different propagation characteristics and vary widely in usefulness. Some frequencies are better than others are for mobile applications. In particular, spectrum between 100 megahertz and 3 gigahertz—the "beachfront property"—is increasingly valuable. All of this beachfront spectrum has already been allocated.

New spectrum technologies promise to remedy this situation if we can develop policies to accommodate them. However, the pace and timing of their introduction and the interaction of spectrum-sharing devices with existing public safety, cellular telephone, and digital broadcast services pose complex management challenges. Interference parameters for spectrum-sharing technologies must be developed to ensure that critical services are not disrupted. Spectrum-sharing technologies may require us to adopt more exacting standards for receivers and transmitters. Change must be closely tied to research and testing, but once the tests are done, the United States needs a spectrum-management process that can act on the results or we will see technologies invented here first put into use somewhere else.

The existing spectrum-management structure is overwhelmed by technological change and strenuous competition. This combination of new demands and new technologies will only become more difficult, given the continuing pace of innovation. Spectrum management in the United States must change to cope with the new environment. Many other developed countries have already restructured spectrum management. The common features of these restructurings have been to streamline agencies, reduce the role of government, allow greater use of markets, and develop national spectrum plans. The United States, however, has not changed. This is not the fault of any agency or person, but the result of a process that blocks innovation. A broad range of commentators now point to the problem this creates and call for a new approach to U.S. spectrum management.

Recommendations to Improve Spectrum Management

In 1934, when the Communications Act became law, Congress and the White House did not want a spectrum czar. As a result, spectrum management is divided between two agencies. The Federal Communications Commission (FCC)—an independent regulatory body that reports to Congress, not the executive branch—has authority over commercial and nonfederal spectrum use. The Commerce Department's National Telecommunications and Information Administration (NTIA) has authority over federal spectrum use. In recent years, the two agencies have worked well together, but our concern is that the growing difficulties of spectrum management will overwhelm this divided process and complicate difficult decisions regarding safety, security, and economic growth.

The United States can improve dispute resolution, accommodate new technologies, reallocate spectrum to more beneficial uses, and safeguard important existing services. To better use a valuable resource, the CSIS commission recommends the following:

- Development by the White House of a comprehensive national strategy for spectrum that addresses economic and security issues and creates a roadmap for change;
- Establishment of a senior White House position for spectrum management and a senior-level Policy Coordinating Committee to resolve disputes

among agencies, interpret and implement policy, and ensure coordination and responsiveness;
- Concentration of responsibility for spectrum-related international activities, including the World Radiocommunication Conference, in a well-resourced ambassadorial position at the State Department;
- Creation of a White House advisory group for national spectrum issues;
- Setup of a public/private research consortium for spectrum research, to lay out a roadmap for U.S. research and development in wireless technologies and promote their adoption and exploitation—which consortium could also provide independent assessments of spectrum issues to support the White House.

Recommendation 1. White House Oversight

The range of participants and issues involved in spectrum management—including national security, economic, diplomatic, and public safety—would leave any agency hard pressed to assert authority. Only the White House has the authority needed to resolve interagency disputes among widely disparate agencies. The White House staff, responsible for supporting the president in security and economic issues, would best perform the task of coordinating issues and stakeholders in spectrum management. For spectrum management, this requires creating a new special assistant to the president for spectrum management and establishing an interagency Policy Coordinating Committee.

Special Assistant for Spectrum Management

The special assistant should be a joint position at both the National Security Council (NSC) and the National Economic Council (NEC), given the major implications of spectrum management for both economic and security issues. The NSC and NEC provide a mechanism to manage problems, ensure broad oversight and continuity, and resolve disputes that is unmatched by other parts of government. The special assistant for spectrum management will have three primary responsibilities:

- Oversee for the president the development and implementation of a national spectrum strategy;
- Chair a new senior interagency group for spectrum management that would develop the national strategy and serve as a dispute resolution mechanism for interagency spectrum issues;
- Provide guidance, continuity, and interagency coordination for U.S. policy objectives in international spectrum negotiations.

Policy Coordinating Committee for Spectrum Management

In addition, we recommend the complementary step of creating a Policy Coordinating Committee (PCC) for spectrum management. A new, senior interagency spectrum group should draw upon senior representatives from relevant agencies (FCC, NTIA, DOD, the new Department of Homeland Security, and other agencies). This group would advise and assist the president on spectrum policy, resolve disputes, and serve as a mechanism for coordinating policies among government entities.

PCCs provide for policy coordination among agencies, provide policy analysis for senior administration decisionmakers, and ensure timely responses to decisions. Among the PCC's most important tasks would be dispute resolution. Creating an interagency dispute resolution process would eliminate many serious spectrum-management problems faced by the United States. It also has the advantage of not requiring changes to NTIA and FCC authorities. This White House group would, like other PCCs, decide interagency disputes or escalate them to the cabinet level or the president for decision.

The president has the authority to adjudicate disputes between cabinet agencies. He does not have the same authority over the FCC, an independent regulatory body. Although this removes the commission from direct presidential oversight, it does not rule out close coordination. The Federal Reserve Board, for example, is an agency "independent within the government" that works closely with the Treasury Department to develop and implement economic and monetary policies. The United States can manage spectrum by using a similar combination of independence and coordination.

Recommendation 2. Spectrum Advisory Board

We also recommend that the White House create a small, high-level advisory group for spectrum, composed of members selected from outside of the government. Advisory boards offer the president authoritative knowledge and insight not otherwise available on key national issues. Spectrum management has now become this sort of issue. The primary responsibilities of a Spectrum Advisory Board would be to:

- Annually assess the effectiveness of the composition and structure of spectrum regulation and make recommendations for improvement or change;
- Serve as a resource for developing long-term spectrum policies;
- Provide advice on weaknesses or deficiencies in spectrum policy and help focus agencies on future challenges.

The new Advisory Board would not have a management role. Its most important function would be to provide an impartial assessment of the interagency spectrum process. While the bifurcated interagency process currently works well, it is not ideal. We have not recommended eliminating or combining agencies, but the board would advise the president if or when this became necessary.

Recommendation 3. Reinforce International Functions

The United States needs to treat international spectrum negotiations more seriously. There is little disagreement that the government could improve its handling of international spectrum management. International coordination of spectrum allocations is increasingly important as telecommunications and wireless markets become global—and as the United States emphasizes the use of sensors and communications technologies for its global military presence. Increased commercial applications also mean that the nation faces economic challenges because of spectrum allocation decisions.

The International Telecommunications Union's World Radiocommunications Conference (WRC) is the most important international spectrum negotiation. WRC negotiations are shaped not only by technical requirements and commercial interests, but also by external political events that can complicate the task of the U.S. delegation. These are complex negotiations where the United States, which has only a single vote, must win the support of many other nations (who are often organized into regional blocs) to protect and advance its interests. Perseverance, resources, and an early start are crucial to success to provide time for the United States to win other nations' support before regional blocs have locked into positions on the various issues.

The United States appoints an ambassador a few months before the WRC begins to conduct negotiations. The position lasts only six months to avoid the need for Senate confirmation. The nation has been fortunate in its choice of ambassadors to the WRC, but an appointment late in the WRC cycle means they often must play catch-up with their foreign counterparts. We recommend that United States reinforce its negotiating efforts by the early appointment and confirmation of a WRC ambassador and by placing the preparation of spectrum negotiations under White House purview.

The State Department's Office of Communications and Information Policy (CIP) is led by the U.S. coordinator for international communication and information policy. The incumbent holds the rank of deputy assistant secretary and is often an ambassador. Incumbents have performed well, but there are no benefits to having two ambassadors. Our recommendation is to merge the two positions into a single, full-time, political-appointee position (the ambassadorship should not be made a career position) and for the president to appoint this ambassador at least one year before the start of the WRC to serve for the duration of an administration.

Creation of a new NSC/NEC special adviser and the Spectrum PCC will also reinforce U.S. efforts internationally and help ensure adequate support for the ambassador. The White House should demonstrate the importance of the new position by seeking amendments to the State Department's authorizing legislation to permanently establish and fund a senior ambassadorial position for spectrum negotiations.

Recommendation 4. Research Support for Spectrum Innovation

The fourth recommendation is to establish a new research consortium for spectrum, supported by both government and private sources. This is in some ways the most radical of the recommendations. We make it because of mounting evidence that research in the United States is declining, while it is continuing to increase overseas. This trend will damage U.S. economic competitiveness and security if not reversed. We also make this recommendation because of the potential for new technologies to allow for more intensive use of spectrum and overcome spectrum "shortages." Technological innovation is the only long-term solution for spectrum access. An investment in research and development will make spectrum management easier.

Absent federal intervention, the United States may not make this investment. In the last decade, the bulk of the funding of research and development (R&D) has shifted to the private sector. Intense global economic competition means that current private-sector R&D in the United States focuses more on development of new products than on research. In contrast, foreign competitors in Europe and Asia gain an advantage from government funding for both short- and long-term basic research.

A new consortium could be organized independently or under the aegis of the National Science Foundation. By bringing leading technologists and managers together for a few months to a year or more, it would provide a resource to the government, industry, and universities for technical issues. This work should initially involve only U.S. scientists and engineers but should in the future expand to an international effort with the United States playing a leadership role.

The new consortium would be a focal point for establishing goals for technology development. It could develop and continually update a technology roadmap that would identify major research areas for spectrum. It could help identify the basic research needed for spectrum innovation (including longer-range research by universities) and participate in performing that research. The consortium could sponsor research in advanced technologies and develop new metrics for interference. The consortium would be an independent and neutral platform for testing potential conflicts between spectrum-using devices or architectures and for the development of standards. These are essential activities for increasing the efficient use of spectrum that the private sector may not adequately fund.

The changing pattern of U.S. R&D funding and the challenge of foreign competition create a long-term risk for the United States. The research consortium's mission would be to reverse this trend. Well-designed U.S. support for research, which does not duplicate or replace private-sector efforts and which involves minimal intervention in private-sector decisionmaking, could enhance U.S. research in spectrum technologies.

Recommendation 5. Develop a National Spectrum Strategy

We join a number of studies on spectrum policy in calling for a national spectrum strategy. A national strategy was not necessary when there were fewer uses competing for spectrum and the technologies that used it were relatively homogenous, but this is no longer the case. Developing a strategy will be difficult in an era of commercial uncertainty and technological change, particularly with the highly diverse and competitive communities that have an interest in spectrum matters.

We propose that the strategy consider and prioritize national spectrum-management goals and identify the policies to achieve them. Creation of the spectrum-management strategy should be the responsibility of the new White House structure we have recommended. A senior advisory broad and a spectrum R&D consortium could support the Spectrum PCC in developing a strategy.

In calling for a national spectrum strategy, we are not calling for central planning. A national strategy that sought to impose a centrally planned approach for spectrum use or that attempted to control spectrum allocation would make matters worse, not better. Strategy is not a pseudonym for economic planning or industrial policy. At the same time, an ad hoc or reactive approach no longer works for spectrum management. The United States cannot rely on market forces to achieve an optimal outcome for spectrum, and a national strategy will confront a series of specific issues. These include 3G, Wi-Fi, ultra-wide band, and digital broadcasting. Beyond these specifics, a few broad issues will shape a national spectrum-management strategy. A national strategy will need to:

- Balance private- and public-sector spectrum needs;
- Determine where the national interest is best served by markets and expanded property rights, by a "commons" model or by continued government control;
- Establish the pace and timing of the introduction of innovative wireless technologies;
- Protect safety-of-life services.

Each raises a series of difficult subsidiary issues, including how to meet new demands while minimizing disruption to existing services; encourage more efficient use of spectrum by both government and private-sector users; clarify incumbent rights; mesh national priorities and international negotiations; promote innovation; and decide where further deregulation is appropriate. It may take several iterations of a national strategy to work through these problems. This should not deter the effort. The national strategy should be a process for planning that establishes a regular cycle of review and revision for U.S. spectrum management.

Conclusion

Spectrum management falls in a special class of political problem that is created by technological change. While technological innovation is the only long-term solution to physical constraints in the supply of spectrum, the existing management structure slows or blocks innovation. Reaping the benefit of new technologies requires reorganization, but reorganization is difficult. The objective in making these recommendations has been to focus on pragmatic, achievable goals to streamline the process for decision and reinforce consideration of broad national interests, so that the United Stats can begin to make the changes needed to gain the full benefits of this immense economic resource.

Introduction

U.S. spectrum management has not kept pace with changes in technology and spectrum use. Explosive demand for mobile information services and new wireless technologies has led to intense competition for access to radio spectrum from both the public and private sectors. While spectrum's value and uses have changed, the United States continues to manage this essential resource with a legal and management process that dates to the beginning of the twentieth century. U.S. spectrum management is obsolete.

Radio spectrum is an essential resource for security and economic growth. Mobile, digital, wireless technologies are the next phase in the transition to an information economy that began 20 years ago. These technologies increase our ability to obtain, process, and exploit information. This makes them a source of competitive advantage, and their use is reshaping business, economic activity, and military operations. Innovative technologies—if they can be successfully deployed—also provide an indispensable opportunity to manage the rising tide of demand for a resource whose supply is fixed.

The physical reality is that the spectrum cannot be expanded. The engineering reality is that we can devise new technologies to use it more efficiently. While many say that we should move forward with caution, most agree that we must introduce new technologies and provide opportunities to test and innovate as demand for spectrum continues to build. We have no choice but to take advantage of the opportunities offered by new spectrum-using technologies.

The pace and timing of how to move ahead are at issue. Spectrum management has become a conflict between technological change and outmoded governance. Advances in technology are creating new wireless equipment and services. In turn, these have led to intense competition and demand for radio spectrum. However, the very technology that has created exploding demand under the existing system of spectrum management could also provide the opportunity to meet this demand, if we had a legal and management structure that could take full advantage of this opportunity.

The current spectrum-management structure hobbles our ability to gain the full benefit of wireless technologies. It stands in the way of change and works against lasting solutions. This is not the fault of any agency or person, but the result of using an outmoded legal and regulatory structure to manage increasingly complex issues. The Center for Strategic and International Studies (CSIS) established a commission to assess the status of spectrum management and consider changes in policies and procedures that would strengthen national security (see pages 41–43 for a list of commission members). The goal was to find practical policy recommendations to replace piecemeal decisionmaking driven by constituent interests with a strategic process oriented towards meeting defined national goals. This report grows out of the work of these experts and provides an overview of the issue and recommendations on how to move ahead to solve four key problems that face U.S. spectrum management. These are:

- The lack of a single effective mechanism for resolving disputes among government agencies over spectrum use;
- The need for a long-range plan or vision for spectrum use to guide policy and provide a greater degree of certainty for investors and greater clarity for innovators;
- The risks to U.S. security and economic growth from a potential lag in the development and use of new technologies;
- The requirement to be better organized to meet the increasing challenges in international spectrum negotiations.

The very technology that has created the mismatch between exploding demand for access to the spectrum and the inability to satisfy that demand under the existing institutional arrangements provides the mechanism to resolve the problem.

The advance of technology has led to steadily expanding uses of the spectrum—and, therefore, an explosion of demand, some realized, some potential. It is that unrealized, potential demand that is the basis of improved communications and further economic growth. The existing organization and legal structure, inherited from an earlier era, blocks the effective utilization of technology—and thereby impedes the improvement in communications and economic growth.

The prospective solution is there. The technology that is now blocked could be exploited to expand the effective utilization of the fixed spectrum—and to provide the additional supply to satisfy growing demand. But to exploit these technologies will require a new framework for management.

The Need for Change

The nations that best manage spectrum will have an advantage over others. Access to radio spectrum has become essential for national security, public safety, and economic growth. Public policy decisions on spectrum will help determine the technological, economic, and military leaders of the twenty-first century. Expanded use of wireless devices offers the U.S. greater security and stronger economic growth. The United States could put its leadership in this vital area at risk if it manages spectrum less effectively than other countries.

We would not be the first country to restructure spectrum management to make more efficient use of this resource.[1] Other developed countries have streamlined spectrum-management agencies, created national spectrum plans, reduced the role of government, and made greater use of markets. In contrast, the United States has not changed its basic structure for spectrum management since

[1] Johannes M. Bauer, "A Comparative Analysis of Spectrum Management Regimes," paper presented at the 30th Communications and Internet Research Conference, Alexandria, Virginia, September 28–30, 2002.

it was created during the New Deal.[2] There is broad agreement that it is time to change. For example:

- President Bush, in his June 2003 memorandum announcing a broad review of federal spectrum use, said, "The existing legal and policy framework for spectrum management has not kept pace with the dramatic changes in technology and spectrum use."[3]
- The Defense Science Board in a review of spectrum needs for national security stated, "our current national governance structure...cannot consider all demands for spectrum and determine which allocations are in the best overall national interest."[4]
- Cellular Telecommunications and Internet Association president Thomas Wheeler said in congressional testimony, "Time is running out for us to make decisions on long-term spectrum management...we need to create a more efficient spectrum management process that focuses more on policy goals than on constituent interests."[5]
- A Toffler Associates report found that "in place of a plan, the US has made a series of ad hoc decisions...The outcomes are a crisis management approach when new services or innovations emerge, artificially spiraling prices...and a disincentive for wireless communications firms and other interested parties to plan or make any investments beyond the next spectrum fight."[6]
- The Federal Communications Commission's Spectrum Policy Task Force reported in November 2002, "Increasing demand for spectrum-based services and devices is straining longstanding and outmoded spectrum policies."[7]
- The General Accounting Office, in a January 2003 study, found that "under the current framework for managing spectrum, it has been difficult to resolve conflicts...Moreover, in the current regulatory environment, no one agency has been given ultimate decision making authority over all spectrum in the United States or the authority to impose fundamental reform...."[8]

---

[2] The 1934 Communications Act created the Federal Communications Commission and gave it responsibility for spectrum regulatory functions previously exercised by the Commerce Department. Commerce retained coordination and management functions for federal spectrum users. The 1934 Act retained elements of earlier laws that dated back to 1912.

[3] The White House, "Fact Sheet on Spectrum Management," June 2003.

[4] Defense Science Board Task Force on DoD Frequency Spectrum Issues, *Coping with Change: Managing RF Spectrum to Meet DoD Needs* (Washington, D.C.: U.S. Department of Defense, November 2000).

[5] Testimony of Thomas E. Wheeler before the Senate Commerce Committee, July 31, 2001.

[6] Steven Kenney, John O'Connor, Richard Szafranski, *Creating the Future of Spectrum Allocation* (Manchester, Mass: Toffler Associates, September 2001).

[7] Federal Communications Commission (FCC) Spectrum Policy Task Force, *Report of the Spectrum Efficiency Working Group* (Washington, D.C. Federal Communications Commission, November 2002).

[8] General Accounting Office, *Comprehensive Review of U.S. Spectrum Management with Broad Stakeholder Involvement Is Needed* (Washington, D.C.: GAO, January 2003).

In the past, the United States could afford a less organized approach to spectrum decisionmaking. Four trends make a new approach necessary. First, U.S. military spectrum requirements have increased as military strategy emphasizes information dominance. Second, increased demand and new technologies have created intense competition among current and potential spectrum users. Third, the sheer scope and size of the spectrum-using market overwhelms the existing management structure. Finally, some new technologies use the spectrum in ways that do not fit well with the command-and-control model for spectrum management originally designed for analog radio broadcasting.[9]

These trends also shape the political landscape for spectrum management. Spectrum-using devices are becoming commodities, and consumers are graduating from passive radio and TV receivers to new devices that transmit radio signals as well as receive them.[10] The creation of a mass market for wireless devices and services will make spectrum a political issue, as spectrum-management decisions will now affect the interests of millions of voters in a new way. They will be increasingly intolerant of policies that do not meet their needs.

A National Resource

The ability to use the radio spectrum plays a crucial role in both security and commerce. Growing demand for access to information has raised the value of spectrum access to all users, and demand for access continues to increase in key sectors of our society. First, access to radio frequency spectrum is essential for military operations and training and for national technical means of intelligence collection. Mobile, wireless applications provide the information infrastructure needed for military advantage in the twenty-first century. The Department of Defense (DOD) has more than 800,000 radio frequency emitters worth over $100 billion, and DOD's operational requirements are growing at 18 percent per year.[11] The Department of Defense faces increasing challenges in meeting its domestic and international needs for spectrum access. In the current environment, marked by the war on terrorism, that access is indispensable for national security.

Military advantage increasingly comes from information superiority, and information superiority depends on access to spectrum. Military spectrum requirements, like commercial requirements, are growing rapidly. The new information-intensive mode of war fighting developed by the United States relies on access to spectrum. The volume of communications continues to increase; operations in Kosovo required 10 times more bandwidth than was used in the Gulf War even though the number of personnel deployed in Kosovo was only one-tenth the number deployed to the Gulf.[12] Operations in Afghanistan and Iraq

[9] The FCC's Spectrum Policy Task Force recognized that existing services such as public safety and broadcasting will continue to be subject to the current regulatory regime. The current interference protections for existing public safety, cellular telephone, and broadcast services remain critically important for these services to function properly. FCC Spectrum Policy Task Force, *Report of the Spectrum Efficiency Working Group*, p. 6.

[10] James Schlesinger conversation with FCC chairman Michael Powell, July 9, 2002.

[11] Defense Science Board Task Force, *Coping with Change.*

[12] *New York Times*, April 16, 2001, p A32.

required more than 40 times the spectrum needed in 1990. The more flexible and precise military force that the United States is developing to use mobile networks and sensors will have a real advantage over potential opponents if we can provide spectrum resources.

Access to radio spectrum also plays a crucial role in public safety and homeland security. First responders depend on tactical radio systems. The events of September 11, 2001, found many fire and police departments unable to communicate with each other by radio. Greater connectivity for first responders and expanded access and reliability for public safety networks is crucial for homeland security. Dissemination of news and information by broadcast television and radio plays an important role. Safety of flight, which depends on both radio communication between cockpit and ground and on navigation systems that often use very low-powered signals, also needs access to spectrum that is safe from interference.

Industries worth hundreds of billions of dollars depend on spectrum, and new industries will continue to appear as costs for new wireless technologies continue to drop in price.[13] Data from the International Telecommunications Union (ITU), the organization responsible for international spectrum coordination, shows that for mobile cellular telephony alone, the number of subscribers globally went from 16 million in 1991 to 955 million in 2001. The United States saw an increase from 7 million to 128 million subscribers in the same period.[14] Sales of "Wi-Fi" wireless networking products quadrupled in 2002 and continue to grow. Thousands of "open access" wireless network connections are being put in place in the United States using Wi-Fi technologies, and the number of Wi-Fi access points is expected to reach 118,000 by 2005.[15] The ITU notes that "40 different kinds of radio services now compete for spectrum allocations to provide the bandwidth needed to extend services or support larger numbers of users."[16]

The evolution of wireless communications will offer a range of new services that will generate intense consumer demand and increase economic productivity. New and more intensive applications have the potential to match the spectacular growth rates of the last decade for cellular telephony, e-mail, or Web browsing and could provide for continuing increases in productivity. The potential economic benefits of spectrum access are immense and will be a key source of

---

[13] Cellular Telecommunication and Internet Association's (CTIA) *Semi-Annual Wireless Industry Survey* for December 2002 shows $80 billion in revenue and $126.9 billion in cumulative capital investments for the U.S. wireless industry. Cellular Telecommunication and Internet Association, *Semi-Annual Wireless Industry Survey* (Washington, D.C.: CTIA, December 2002).

[14] International Telecommunications Union (ITU), "Key Global Telecom Indicators for the World Telecommunication Service Sector," 2001, http://www.itu.int/ITU-D/ict/statistics/at_glance/KeyTelecom99.html; Cellular Telecommunication and Internet Association, *Semi-Annual Wireless Industry Survey* (Washington, D.C.: CTIA, 2003). Most estimates put the number of mobile phone users at over 1 billion by 2002.

[15] Pui-Wing Tam and Nick Wingfield, "Stereos, Hand-Helds, TVs Are Now Going Wireless," *Wall Street Journal*, April 23, 2003; Yuki Noguchi, "Wi-fi Networks are Expanding Internet's Reach, Profit Opportunities," *Washington Post*, April 20, 2003, p. H1.

[16] ITU, "Overview of the ITU and Its Three Sectors," http://www.itu.int/aboutitu/overview/o-r.html.

future economic growth if the United States can organize itself to take advantage of them.

As the pace of technological innovation increases, and as the rewards from these innovations accrue to the nations that can adopt them before others, current U.S. spectrum management puts us at a disadvantage and threatens to make that disadvantage greater. The adoption of cellular telephony illustrates the sort of problem the U.S. will face if it does not improve spectrum management. Although a leader in cellular innovation, the U.S. was slow to approve the use of cellular telephony. The introduction of cellular technologies and the valuable services they provided was delayed for years. One study put the loss to the U.S. economy from delays in approving the use of cellular communications at $86 billion.[17] Another study concluded that the delay "significantly diminished American industry's initial lead in cellular radio technology."[18] Our concern is that despite considerable effort and improvement by the regulatory agencies, the way that the United States makes spectrum policy decisions creates considerable risk for similar delay and loss in the future.

New Spectrum Technologies Create Opportunity and Risk

Wireless technologies have entered a new era. The last two decades have seen a series of technologies emerge that fundamentally change how the radio spectrum can be used. Innovative spectrum-based technologies are now being developed at a rapid pace. These technologies offer the possibility of valuable new services, increases in productivity, and the potential for a much more intensive use of spectrum.

But new services and technologies pose significant challenges for spectrum management, because the way they use spectrum does not fit well with the current U.S. approach to spectrum management. The rationale for the existing approach is the mitigation of interference. Powerful, unregulated transmitters that blotted out weaker stations marked the early days of radio. Since 1912, when many believed that interference blocked distress signals from RMS *Titanic* sinking in the North Atlantic, the United States has treated spectrum as a publicly owned resource whose use requires federal approval.[19] The licensing process controlled interference by regulating transmitters and by assigning specific rights to a unique frequency to different users or classes or users.

Most wireless technologies developed before 1990 transmit and receive signals using a single, narrow frequency. This made it easy to allocate spectrum exclusively to individual operators who used a specified frequency at a specified power in a specific geographic area. Transmitters that enter the spectrum territory

[17] J. Rohlfs, C. L. Jackson, and T. E. Kelly, "Estimate of the Loss to the United States Caused by the FCC's Delays in Licensing Cellular Telecommunications," National Economic Research Associates Report, 1991.

[18] Dale Hatfield, *Spectrum Issues for the 1990s: New Challenges for Spectrum Management* (Washington, D.C.: Annenberg Washington Program in Communications Policy Studies, Northwestern University, 1995).

[19] With the passage of the Federal Radio Act of 1912, which first created federal regulation of spectrum.

of a licensed user and cause interference with the approved signal are forced to stop transmitting. Frequency assignments often include "guard bands," empty blocks of spectrum on either side of an assigned frequency, intentionally kept unused in order to avoid interference. As spectrum was increasingly divided among licensees, this exclusive-use approach made it difficult for new entrants or technologies to gain access to spectrum and resulted in shortages, even though recent studies have found that large parts of the spectrum are unused most of the time.

Many new spectrum technologies do not fit this regulatory paradigm of exclusive use. Powerful microprocessors, software, digital signal processing, and improved antennae have created new ways to transmit and receive radio signals. They allow wireless devices to move beyond the traditional spectrum differentiators of location, frequency, and code. New technologies exploit gaps in transmission and differences in time of use to allow a more intensive use of the radio spectrum.

Spread spectrum, software-defined radio (SDR),[20] and ultra wideband are examples of the new technologies. Spread-spectrum technologies deliberately "hop" among a large number of frequencies for very short periods of time. The popular 802.11 Wi-Fi wireless local area network system uses spread spectrum. In contrast to spread spectrum, ultra-wideband technologies transmit signals simultaneously over a broad range of frequencies. The signal's low power, advocates say, prevents interference with other devices operating in the same frequencies. Software-defined radio uses computer chips to control transmitters and receivers. Programs run on SDR devices allow them to send or receive signals using many different frequencies and transmission protocols, allowing a more flexible use of spectrum.

Often, the new technologies grow out of work originally done for the Department of Defense. Military radio technologies emphasized mobility and resistance to interference to avoid interception or jamming by opponents. Mobility and resistance to interference are commercially desirable attributes. While some technologies are experimental, others (such as Wi-Fi) have become mass-market commodities sold in the millions. The common features of the new technologies are the ability to pack more signals and data into the same spectrum, to dynamically adjust their use of spectrum, and to reduce interference through technical capabilities rather than by exclusive assignments and guard bands, allowing many devices to use the same frequencies.

However, the assertion that new technologies can share spectrum with existing uses has raised significant concerns among incumbent spectrum users. Tests by incumbents during the debate over the FCC's new ultra-wideband regulations suggested that essential services like global positioning system (GPS) or aircraft landing systems would be disrupted. Other tests suggested that more cell phone calls would be dropped.[21] The National Aeronautics and Space Administration

---

[20] SDR that run programs that are more sophisticated are sometimes known as "cognitive radio."

[21] Erika Jonietz, "Ultrawideband Squeezes In," *Technology Review*, September 2002.

(NASA) says its operations would face interference. Greater use of new wireless technologies may require new regulations requiring manufacturers to build better receivers and transmitters, as many existing devices (like television sets) do a poor job of handling interference. This will increase the costs (at least initially) of wireless devices. Introducing new wireless technologies also requires the United States to think about and measure interference in new ways and to define what interference levels are acceptable and what levels are damaging.

Incumbent spectrum holders in both the public and private sectors want to move very cautiously to ensure that innovations do not disrupt valuable, existing services.[22]

These concerns create conflict and debate among incumbents and new claimants. Although exclusive-use licensing may for the foreseeable future still be appropriate for many existing public safety, cellular, and broadcast operations, new wireless technologies do not fit with a regulatory approach designed and put in place decades ago for very different markets and equipment. The potential interaction of spectrum-sharing devices, licensed or unlicensed, with vital services poses a complex management challenge, and the mixture of existing allocations and new demands creates intense competition and drives the need for better spectrum management.

The physical characteristics of the radio spectrum also increase competition. Different parts of the spectrum have different propagation characteristics and vary widely in their usefulness and value for different applications or services. The use of some frequencies requires very large antennae, and some frequencies are better than others for mobile applications. In particular, spectrum between 100 megahertz and 3 gigahertz—which we call the "beachfront property" for the allocation of spectrum—has become increasingly valuable. Devices that operate in this frequency range can use smaller antennae and lower-power transmitters and are less vulnerable to atmospheric conditions, making them more appealing to businesses and consumers. This beachfront spectrum has already all been allocated. The problems of a crowded beachfront are compounded when international allocations assign spectrum to new commercial services—and it is already used by the United States for military or other purposes (as occurred with spectrum allocated internationally for "3G" wireless services and for 5 Ghz unlicensed services).

Emerging spectrum technologies hold tremendous promise to remedy this situation if we can develop policies to accommodate them. They could provide a substantial opportunity to balance supply and demand for scarce radio spectrum because of their ability to pack more traffic into a given amount of spectrum than can previous wireless technologies. However, innovators find it difficult to develop, test, and bring to market devices and services based on new technologies. Some research and development efforts have been driven to using amateur radio licenses to test new systems. Others "illegally" (i.e., without FCC approval) test

---

[22] Interference issues may be minimized when spectrum has been set aside and devoted exclusively for use by new technologies, but problems may y arise when these devices share spectrum with incumbent uses licensed under the existing regulatory regime.

their devices in unpopulated areas.[23] It will be increasingly difficult for the United States to take advantage of innovation, despite its rapid pace, unless the current approach to spectrum management changes. Deploying innovative technologies must be closely tied to research and testing, but once the tests are done, the nation will need a spectrum-management process that can act on the results, or we will see technologies invented here and first put into use somewhere else.

Recommendations for Improved Spectrum Management

These technological and economic developments in spectrum impinge on and compete with each other, but there is no effective process for setting goals or resolving conflicts. The roots of this problem lie in the U.S. approach to federal government, which relies on the dispersion of power. The Constitution intentionally divides authority among the Congress, the executive, and the judiciary, and between the federal government and the states. Fragmented authorities prevent a powerful sovereign from overwhelming elected government.

This division of authorities carries over into agency responsibilities. Unlike other countries, spectrum management in the United States is fragmented. In 1934, when the Communications Act became law, Congress and the White House agreed that they did not want a single spectrum czar. Since that time, two agencies have managed the use of spectrum. The Federal Communications Commission (FCC), a regulatory body that reports to Congress, not the executive branch, manages commercial and non-federal spectrum use. The Commerce Department (specifically the National Telecommunications and Information Administration—NTIA) manages federal government spectrum use. Other agency stakeholders also play a role in spectrum management. The State Department oversees international spectrum negotiations. The Defense Department, the largest government user of spectrum, plays a key role in reallocation issues. NASA and the Federal Aviation Administration (FAA) have important spectrum equities based on their agency missions and safety-of-flight requirements.

With this structure, it is natural to be concerned with conflict. In recent years, the FCC and NTIA have worked well together, but the growing challenges of spectrum management will overwhelm this fragmented process in the future. The nation needs a better system to make the difficult trade-offs among safety, security, and economic growth now required for national and international spectrum management.

A lack of planning compounds the organizational challenges in spectrum allocation and management. New technologies create the possibility of expanding spectrum use, but they may also pose potential risks to existing wireless services that are both important and valuable, such as public safety, broadcasting, and safety of flight. There is reasonable concern that the current regulatory structures and management processes will become an obstacle to change. Organizational issues also compound U.S. difficulties in international coordination of spectrum access, which has become vitally important both for military and economic reasons.

---

[23] National Science Foundation, "The Future of Spectrum Workshop," May 28–29, 2003.

Technology compels us to seek a new approach that can capture the advantages of expanding innovation rather than frustrating them. In an era of rapid technological change, increased demand, and intense competition, the United States needs a new approach to spectrum management. The goal is to improve dispute resolution among the federal agencies, speed decisions on new technologies and reallocation of spectrum, and better safeguard important existing services. To make better use of a valuable resource, the CSIS commission has produced recommendations to replace piecemeal decisionmaking with a strategic process oriented toward meeting long-term national goals and a management system designed for the information economy. Our specific recommendations:

- Develop a comprehensive national strategy from the White House for spectrum policy that addresses economic and security issues and creates a roadmap for change;
- Establish a senior White House position for spectrum management and a senior-level Policy Coordinating Committee to develop policy, ensure coordination and responsiveness, and resolve disputes among agencies or between agencies and the FCC;
- Concentrate responsibility for spectrum-related international activities including the World Radiocommunication Conference in a permanent, well-resourced ambassadorial position at the State Department;
- Create a White House advisory group for national spectrum issues;
- Set up a public/private research institution for spectrum that would expand U.S. research and development efforts in wireless technologies, provide independent assessments of spectrum use and the implications of new technologies, and support the White House and the Senior Advisory Group in making policy.

R E C O M M E N D A T I O N 1

White House Oversight

Perhaps the most persistent pattern that recurs over the last one hundred years of efforts to reorganize the executive branch of the Federal government is fierce resistance.—Making Democracy Work: A Brief History of 20th Century Federal Executive Reorganization

Special Assistant for Spectrum Management

The broad range of participants and issues involved in spectrum management—including national security, economics, diplomacy, and public safety—would leave any single agency hard pressed to assert authority. Only the White House has the authority to resolve interagency disputes among such widely disparate departments. The White House staff that supports the president is best suited to determine national policy when conflicts arise in spectrum management.[1] We recommend that the White House name a special assistant to the president to oversee spectrum management and establish a new interagency Policy Coordinating Committee. Creating these through an executive order would give the new structure the appropriate authority and weight. The executive order should assign the new special assistant for spectrum management position three primary responsibilities:

- Oversee for the president the development and implementation of a national spectrum strategy;
- Manage a new senior group for spectrum management that would develop the national strategy and serve as a dispute resolution mechanism for interagency spectrum issues;
- Provide guidance and continuity and assure interagency coordination and broad adherence to U.S. policy objectives in international spectrum negotiations.

The new special assistant should be a joint position at both the National Security Council (NSC) and the National Economic Council (NEC), given the major implications of spectrum management for both economics and security. The NSC is uniquely situated in the federal government and has the authority and the position to manage problems, ensure broad oversight and continuity, and resolve disputes. It provides the president with advice on the integration of

[1] This recommendation builds on the Defense Science Board's *Coping with Change*, which called for the creation of a new Office of Information Resource Policy at the White House. We believe that a joint NSC/NEC appointment makes a better starting point for a White House role while reserving the option of creating a new office later if this seems warranted.

domestic, foreign, and military policies as they relate to U.S. national security. The NEC plays a similar (if less influential) role for economic issues. Its tasks are to coordinate policymaking for both domestic and international economic issues, ensure that policy decisions are consistent with the president's economic goals, and monitor the implementation of the president's economic policy agenda. One NEC deputy already reports to both the NEC director and the national security adviser, and several NSC special assistants coordinate with and support the NEC director.

A review of the history of federal spectrum management would show that the United States has created on several occasions a White House position to oversee spectrum policy when challenges were great and then eliminated this position when the policy situation was static. The current challenges and the steady increase in the importance of spectrum for both the economy and for security require a renewal of direct presidential oversight of spectrum management.

Spectrum management functions were located at the White House until the Carter administration moved them to the Commerce Department. This decision may have been appropriate at the time, but the increased importance of spectrum and the difficulties of managing it for national purposes have changed this. We do not recommend the creation of an independent office in the White House or the return of the Office of Telecommunications Policy (OTP) from NTIA to the White House. A lean structure, appropriately supported by new interagency and advisory processes, is sufficient.

Policy Coordinating Committee for Spectrum Management

In addition to a new special assistant to the president, we recommend the complementary step of creating a Policy Coordinating Committee (PCC) for spectrum management. A new, senior interagency spectrum group should draw on senior representatives from the relevant agencies (FCC, NTIA, DOD, the new Department of Homeland Security, and other agencies). The functions of this group would be to advise and assist the president on spectrum policies as needed, resolve disputes, and serve as a mechanism for coordinating policy among the several government agencies. One crucial function would be to review underused or unused areas of spectrum and decide whether and how these could be reallocated.

The Commerce Department's Interdepartmental Radio Advisory Committee (IRAC), which dates back to the 1920s, is currently the interagency body that provides for coordination on spectrum matters. The IRAC's relatively large membership ensures that all agencies have a voice in spectrum issues and frequency assignments. We believe that the IRAC should continue to operate as the organization managing federal spectrum use, but that formulating national policies or resolving high-level interagency disputes would be better handled by a White House PCC. The members of a new PCC must be more senior than members of the IRAC.

Interagency coordination between FCC and NTIA is based on a memorandum of understanding (which has been recently revised), but it essentially requires each agency to provide the other with advance notice when it takes a decision that impinges on spectrum that falls under the jurisdiction of both. It does not provide for dispute resolution or planning.

There are now 20 PCCs responsible for regional or functional policy issues and interagency coordination. PCCs provide the forum for the routine coordination of policy issues, provide policy analysis for senior administration decisionmakers, and ensure timely responses to decisions. Either senior agency representatives (with NSC staff acting as an executive secretary for the group) or senior NSC staff chair these groups. The chair has the option of establishing working groups that are subordinate to the PCC to work on specific issues. Many commission members thought that it would be best to begin a new spectrum management process by having the NSC chair the new group, in order to reinforce the centrality of the White House in the policy process and to make clear that this was a break with the past.

Dispute Resolution

Among the PCC's essential tasks would be dispute resolution. This group would, like other PCCs, decide interagency disputes or escalate them to the cabinet level or the president for decision. Creating a unified dispute-resolution process would eliminate many serious spectrum-management problems faced by the nation. Assigning this task to a PCC also has the advantage of requiring little change to NTIA and FCC authorities. Making the new Spectrum PCC responsible for dispute resolution would provide many of the benefits of a single agency and is the least difficult restructuring option to implement.

The president has the authority to adjudicate disputes between cabinet agencies. He does not have the same authority over the FCC, an independent regulatory body. Although this removes the commissioners from direct presidential oversight, it does not rule out close coordination. The Federal Reserve Board, for example, describes itself as an agency that is "independent within the government" and works closely with the Treasury Department to develop national economic and monetary policies. Like FCC commissioners, Federal Reserve Board members are appointed by the president for a fixed term. This helps provides an adequate degree of consistency with the executive branch. The Federal Reserve works within the larger framework of the overall economic and financial policy objectives established by the executive branch. The United States can adopt a similar approach combining independence and coordination for spectrum management.

Although we recommend that the United States first try a "soft-management" approach that provides for greater coordination and focus, one of the tasks of the new PCC would be to advise the White House if this approach was working and to recommend further measures to consolidate spectrum management needed. If cooperation between an independent FCC and a White House–led spectrum-policy structure proves unwieldy, we recommend that the White House seek legislation to subordinate all aspects of spectrum management in the executive branch. This legislation could include a requirement to make the dispute-resolution process subject to congressional notification and approval. Strong congressional oversight is a necessary component for changing spectrum management.

The goal here is to create a single process for spectrum management, replacing the heterogeneous approach now in use. Creating a single group, led by the White House, with oversight for all spectrum issues, responsible for producing a national strategy that applies to spectrum use and buttressed by an annual report to the president on major spectrum issues and the status of interagency efforts to resolve them, will drive the nation to a unified approach to spectrum policy.

RECOMMENDATION 2

Spectrum Advisory Board

We also recommend that the White House create a small, bipartisan, high-level advisory committee for spectrum, composed of members selected from outside the government. Presidential advisory boards bring the president authoritative knowledge and insight not available within the government. The primary responsibilities of a Spectrum Advisory Board would be to:

- Annually assess the effectiveness of U.S. spectrum management in responding to technological imperatives and, as necessary, make recommendations for improvement or change;
- Serve as a forum for the discussion and development of long-term spectrum policy and management needs;
- Provide advice on weaknesses or deficiencies in spectrum policy and help focus agencies on future challenges.

This recommendation draws on and combines a broad range of precedents used by the federal government, including the Base Realignments and Closures Commission[1] and the President's Foreign Intelligence Advisory Broad.

The President's Foreign Intelligence Advisory Board (PFIAB), created in 1956, provides objective, expert advice on intelligence matters. Its primary activities are to assess the adequacy of current intelligence activities, identify future challenges for intelligence, and advise the president on the legality of certain intelligence activities. Although a Spectrum Advisory Board would face a greater challenge than the PFIAB in the sense that intelligence activities do not have the commercial implications of spectrum policy, a careful and transparent process for the selection of board members would ensure a balanced approach.

The Base Realignments and Closures Commission (BRAC) dealt with a range of economic and commercial issues. Congress created the BRAC process in 1988 to handle the politically charged issue of deciding which military facilities to close. The initial round of BRAC decisions successfully avoided politicization. Although later rounds found that local communities and their political representatives had adjusted to the new process and were able to affect it, the process did depoliticize and speed base closings. BRAC identified and prioritized facilities for closure and is a model for transferring resources from incumbents to more efficient uses. A senior spectrum group placed outside of the agencies could take a similar responsibility for promoting efficient use of spectrum allocated to government and the private sector and could help develop a national spectrum strategy. The BRAC process is an example of a useful mechanism for Congress's role in

---

[1] The idea of a BRAC-like body for spectrum management was developed in the Toffler Associates report. See Kenney et al., *Creating the Future of Spectrum Allocation.*

developing a Spectrum Advisory Board. Congress shared in the appointment process in the case of the BRAC. Providing Congress a role in the appointment of a senior advisory group would continue the oversight of spectrum management that Congress now plays.

A number of advisory councils already exist for telecommunications and spectrum matters. The National Security Telecommunications Advisory Committee (NSTAC), created in 1982, provides the president with private-sector views and advice on emergency preparedness and national security communications policy.[2] The FCC has numerous advisory bodies, including committees on technology, consumer affairs, local and state governments, homeland security policy, network reliability and interoperability, and for the World Radiocommunications Conference. In 2002, the FCC established the Media Security Reliability Council (MSRC) to develop recommended policies to insure the proper functioning of broadcast media in the event of a terrorist attack or natural disaster. An NTIA advisory committee, renamed in 1993 as the Spectrum Planning and Policy Advisory Committee (SPAC), due to increased interest in policy matters, has existed since 1965 when it provided advice to the Office of Telecommunications Policy in the White House (before President Carter moved spectrum management functions to the Commerce Department).[3] These agency-level advisory bodies cannot provide the broader overview and advice that spectrum management now requires.

The new Spectrum Advisory Board would not have any management function. Its members, like PFIAB, would serve on a part-time basis in a general advisory function. Its role would be to take a broad view of the needs of both government and the private sector and provide external expertise to the policy process. The board would be an independent source of advice on the effectiveness of U.S. spectrum management and policy. A board that was independent of the many agencies involved in spectrum and that did not have any operational responsibility would be able to provide an outside view on strategy and structure for spectrum management.

The most important function of the new board would be to serve as a watchdog of the interagency spectrum process. Although that interagency process currently works well in most cases, the bifurcated structure creates the possibility of disharmony and inefficiency. We have not recommended the reorganization of spectrum regulatory functions, but we do not rule out the idea that conditions may warrant this change to a single agency in the future. One vehicle for this watchdog function would be to task the board with the preparation of an annual report to assess the effectiveness of U.S. spectrum management, future national spectrum needs, and the policies and that could best meet those needs. This review of interagency effectiveness is an essential function for the "soft management" approach we advocate. An annual report to the president on the state of spectrum

---

[2] See "National Security Telecommunications Advisory Committee (NSTAC)," http://www.ncs.gov/NSTAC/nstac.htm.

[3] See "Spectrum Planning and Policy Advisory Committee (SPAC)," http://www.ntia.doc.gov/osmhome/spac.html.

management would provide an important incentive for agencies to cooperate, in that this report could explicitly recommend whether the situation required changing agency functions or seeking new legislation to create a single agency.

The board could also play a key role in the development of a national spectrum strategy. It will not be an easy task to reconcile the many important national interests that shape spectrum, nor will it be easy to predict the directions that the market and technology may go. These steps, however, are essential for the development of a strategy and will require research, public hearings, and congressional testimony to supplement interagency discussion. The Spectrum Advisory Board could contribute to this process and provide a long-term view of elements and direction for national strategy.

RECOMMENDATION 3

Reinforce International Functions

The U.S. is often not ready to present common positions at international preparatory meetings. This means separate U.S. interests may be negotiating against one another at many meetings. It also means that we lack the influence we need as regional and technical groups are reaching decisions.—Gail Schoettler, U.S. Ambassador to the 2000 World Radiocommunications Conference

There is little disagreement that the United States could improve its handling of the international aspects of spectrum management.[1] International agreement and coordination of spectrum allocations have become increasingly important as telecommunications and wireless markets become global, as the pace of technological change increases, and as the United States emphasizes the use of new sensors and communications technologies for military operations. Increased commercial applications also mean that the United States faces greater economic challenges because of spectrum allocation decisions.

Although the United States is particularly dependent on spectrum access, the division of responsibilities and leadership among the State Department, NTIA, and FCC has precluded the development of a coherent, long-term national strategy for international spectrum negotiations. Career diplomats sometimes see spectrum work as "too technical," and the government has been unwilling to devote the resources needed to assure support for its positions. An approach that depends on a part-time World Radiocommunications Conference (WRC) ambassador, ad-hoc financing, and an intermittent level of activity makes it difficult to organize the support that the nation needs.[2]

The most important international spectrum negotiations, the World Radiocommunications Conference, are conducted for the United States by a temporary ambassador appointed shortly before a WRC begins. The position is temporary and lasts only six months to avoid the need for Senate confirmation. We recommend that the United States reinforce its negotiating efforts by placing the preparations of spectrum negotiation under White House purview and by the

[1] A recent GAO report criticized the short tenure of the WRC ambassador and criticized as inefficient the separate processes the FCC and NTIA use to develop WRC positions. See *Telecommunications: Better Coordination and Enhanced Accountability Needed to Improve Spectrum Management*, GAO-02-906 (Washington, D.C.: GAO, September 2002).

[2] Remarks by Gail Schoettler at CSIS meeting, September 20, 2001.

early appointment and confirmation of a WRC ambassador who leads the State Department office responsible for spectrum negotiations.

The International Telecommunications Union (ITU) is the principal mechanism for international action on spectrum issues. Founded in 1865 to coordinate telegraphy, the ITU is now an agency of the United Nations and serves as the forum and secretariat for the international allocation of spectrum. The ITU's International Table of Frequency Allocations divides the spectrum into frequency bands or blocks and allocates them to broadly defined categories of services. The full set of rules for international spectrum management is found in the ITU's Radio Regulations, which have treaty status and are legally binding once countries ratify them. National spectrum agencies then allocate spectrum use within their jurisdictions, usually in conformance with the ITU allocations.

The only way the Radio Regulations can be revised or updated is through agreement at the WRC held every two or three years and attended by all ITU members. In a period of increasingly global activity and rapid technological change, the WRC has become a crucial arena for spectrum policy. The development of mass-market wireless applications and the increased use of wireless devices by the U.S. military increase the chance that the interests of the United States and other governments will conflict. The Defense Science Board's spectrum report[3] found that many countries are "aggressively asserting" their spectrum rights. Decisions at the WRC could mean that essential U.S. military systems will compete with commercial services when they deploy overseas or that U.S. manufacturers and service providers will be put at a disadvantage.

The United States has been fortunate in that its ambassadors to the WRC have performed well, but upon their appointment they find themselves in a position of having to catch up with their foreign counterparts. Negotiations leading up to the WRC begin several years before the ambassador's appointment. The ITU often schedules the first preparatory meeting for an upcoming WRC shortly after the conclusion of the previous WRC. The latest WRC negotiations provided an example of the disadvantages of a temporary ambassador. The designee was scheduled to speak to an important regional bloc in February 2003, but inadvertent delays in the announcement of her appointment meant that she has to forgo addressing the group.

WRC negotiations are shaped not only by technical requirements and by commercial interests of participants, but also by external political events that complicate U.S. efforts. These are complex negotiations where the United States, which has only a single vote, must win the support of many other nations (who are often organized into regional blocs) to protect and advance its interests. Some of these regional blocs, such as the European Conference of Postal and Telecommunications Administrations (CEPT) use the WRC to internationalize regional decisions and allocations they have already made.[4] Perseverance,

---

[3] Defense Science Board, *Coping with Change.*

[4] See, for example, Germany's Regulatory Authority for Telecommunications and Posts, *Annual Report 2002* (Bonn: Press and Public Affairs Section, 2002), p. 43, which talks of using in the WRC to "push through the European solution for countries outside of Europe," and "replicating at a global level the 5Ghz solution CEPT has found for Europe."

resources, and an early start are crucial to success, to allow the United States to persuade others to support it before regional blocs have locked into positions on the various issues.

The ambassador also joins in midstream the effort by U.S. agencies and the private sector to come up with national positions for the WRC. For the last WRC (June-July 2003), the United States provided its initial 110-page package of proposals in February, before the current WRC ambassador was appointed. The United States develops these positions through a complicated process of consultation among agencies and with the private sector. The FCC coordinates private sector input to the U.S. position. NTIA coordinates the input of federal agencies. The Department of State hosts another group, the International Telecommunications Advisory Committee. To help reduce coordination problems, the current Department of State leadership has convened interagency meetings with representatives from NTIA, DOD, FAA, NASA, and the FCC to reconcile differences in positions and U.S. proposals on conference issues.

A complicated consultative process may be necessary to capture the full range of views on spectrum, but the United States is often not ready to present common positions at international preparatory meetings. At times, according to previous WRC ambassadors, this has meant that separate U.S. interests may be negotiating against one another at some meetings. It also means that U.S. influence is diminished as regional and technical groups are reaching decisions before U.S. positions are even formulated. The ambassador acts as arbiter, and absent the discipline provided by a head of delegation, coordination difficulties and conflicting positions will continue to be a problem. Given the increasing importance of regional blocs and the need to influence them earlier in the process, the United States needs a better way to build support for its proposals at the regional, bilateral, and multilateral meetings that lead up to the WRC.[5] The way to do this is to place the preparations of spectrum negotiation under White House oversight and to appoint a WRC ambassador much earlier in the negotiating cycle.

The State Department, which leads spectrum negotiations, is often blamed for a lack of interest and reluctance to act in a timely manner. It is more accurate to say that problems with the way the United States conducts international spectrum negotiations reflect the fragmented management structure and the historically low priority of spectrum negotiations. One solution would be to assign negotiating responsibilities to another agency. However, previous WRC ambassadors say that despite the ambivalence with which the State Department approaches the WRC, its global presence and political insights are essential for informing and supporting international spectrum negotiating efforts.[6] This argues for keeping the function at the department if it is suitably reinforced by a White House emphasis on international spectrum negotiations as a priority and commitment of the necessary financial resources. The cost is not high, especially when compared to

---

[5] David A. Gross, deputy assistant secretary for international communications and information policy, testimony before the House Government Reform Committee, Subcommittee on National Security, Veterans Affairs, and International Relations, April 23, 2002.

[6] Conversations with former WRC ambassadors Travis Marshall and Gail Schoettler.

the benefits or to the likely harm to our national interests if we do not act decisively.

The State Department's Communications and Information Policy (CIP) Group, located in the Bureau of Economic and Business Affairs, is led by the U.S. coordinator for international communication and information policy. The incumbent holds the rank of deputy assistant secretary and is often made an ambassador. Incumbents in this position have performed well, but the benefits of having two ambassadors (one for CIP and one for the WRC) are unclear. Our recommendation is to merge the two positions into a single, political-appointee position. The ambassadorship should not be a career position. The president should appoint the ambassador at least one year before the start of the WRC, and the ambassador should serve, at the president's pleasure, for the duration of an administration. The early appointment of a long-term ambassador by the president would give the U.S. an effective international presence to achieve its spectrum goals.

Creation of a new NSC/NEC special adviser and the Spectrum PCC will reinforce U.S. efforts internationally and help ensure adequate support for the ambassador. Development of a national spectrum strategy that lays out long-term objectives will also improve international efforts.[7] Given the importance of international spectrum negotiations, the White House may wish to signal the importance it assigns to the new position by seeking amendments to the Department of State's authorizing legislation to permanently establish a senior ambassadorial position for spectrum negotiations. This legislative change would reinforce the importance of the position and spectrum negotiations.

[7] Germany's agency for telecommunications and spectrum, created in 1996, sees a national strategy as a way to drive innovation and to provide "an overview of selected spectrum management areas even before national and international regulatory agencies conduct hearings." See Regulatory Authority for Telecommunications and Posts press release, "Kurth: 'Spectrum management will continue to drive innovation,'" March 18, 2003.

RECOMMENDATION 4

Research Support for Spectrum Innovation

The considerable technical challenges that must be addressed...and the ambitious foreign programs designed to do this, are reminders that continued U.S. leadership cannot be taken for granted.—National Research Council Board on Science, Technology, and Economic Policy

The fourth recommendation is to establish a new research consortium for spectrum, supported by both government and private resources. This is in some ways the most radical of the recommendations. We make it because of increasing evidence that spectrum-related research in the United States is declining, while it is continuing to increase overseas. This trend, if not reversed, will damage U.S. economic competitiveness and security. We also make this recommendation because of the potential for new technologies to make more intensive use of spectrum and overcome "shortages." An investment in exploring these technologies will make the task of spectrum management easier.

Absent federal involvement, adequate investment in research may not be made. In the last decade, the bulk of the funding for research and development has shifted from government sources to the private sector. Intense global economic competition means that private-sector research and development (R&D) in the United States is necessarily focused more on development of new products rather than on research. Only a small number of private R&D centers in the nation still perform basic research to discover new wireless technologies.[1]

In contrast, foreign competitors in Europe and Asia are gaining an advantage from the funding provided by their governments for both short- and long-term research.[2] Europe has established a series of cooperative research programs such as the European Strategic Program of Research and Development in Information Technology (ESPRIT) and Research and Technology Development in Advanced Communications Technologies in Europe (RACE), and some analysts have concluded "The European telecommunications industry has benefited from a series of long-term, large scale European Union–sponsored R&D projects that

[1] Rick Whiting and Aaron Ricadela, "Future Funding: With the Economy in the Dumps, Vendors Struggle to Keep Investing in R&D—and the Future," *Information Week*, October 28, 2002. Whiting and Ricaldela found only three companies' labs still focused on basis research.

[2] Charles Wessner, ed., *Securing the Future: Regional and National Programs to Support the Semiconductor Industry* (Washington, D.C.: National Academies Press, 2003).

began in the 1980s and continue today."[3] Japan and China also have ambitious, government-supported research programs in spectrum-using technologies.

The changing pattern of R&D funding in the United States and the challenge of foreign competition pose a long-term risk for the nation. Over the long term, the result will be to shift spectrum-related innovation outside the United States. While the challenge is clear, a recent National Research Council study notes, "no consensus exists on the appropriate mechanisms or levels of support for research."[4] The proposed research consortium's mission would be to reverse this trend by creating a mechanism that could draw on both public and private support and research efforts.

To succeed, a research consortium would need to clearly identify its goals and establish measurable milestones for spectrum issues. The most important of these would be to create and continually update a technology roadmap to identify major technology intentions and directions for wireless devices. It could also identify the research needed to develop technologies, including longer-range research by universities, and participate in that research. A consortium could assemble best practices for deploying new technologies and dealing with legacy issues. It could provide an independent and neutral platform for the testing of potential conflicts between spectrum-using devices or architectures and develop new metrics for interference. Its mission should include the application of funds for modeling, simulation, and testing to assess mutual compatibility and interference issues. The measurement of interference will play a crucial role in deciding when and where the United States can deploy new spectrum-sharing technologies. These are essential areas for improving the efficient use of spectrum but unlikely to be adequately funded by the private sector.

A new research consortium could also make a useful contribution to the development of standards. Standards development is performed by professional bodies and depends on voluntary efforts.[5] The development of the global system for mobile communications (GSM) standard for mobile telephony illustrates how standards affect competitiveness and innovation. While the history of the GSM standard and its effect on markets is complex,[6] the standard increased non-U.S. market share in the mobile telephony market. Putting aside the technical merits of GSM, and although European standards bodies have been unable to repeat this

---

[3] Thomas Howell, "Competing Programs: Government Support for Microelectronics," in Wessner, *Securing the Future*, p. 240.

[4] Wessner, *Securing the Future*, p. 40.

[5] Standards development for digital broadcasting is challenging because broadcasters have no direct control over the design and manufacture of broadcast receiving devices. The "open architecture" that characterizes broadcast may require greater government coordination and research to facilitate rapid technical development.

[6] GSM faced a "crisis" in 1991 because the slow pace of approval for the new GSM terminals kept networks from using the new standard. The crisis was overcome by using interim equipment approvals based on a shortened review procedure. See "Barriers to Overcome," in *History of GSM* (London: GSM Association, no date), http://www.gsmworld.com/about/history/history_page11.shtml.

success for wireless networks,[7] the development and widespread use of GSM gave European firms an advantage in the global market.[8]

SEMATECH (SEmiconductor MAnufacturing TECHnology) is one model for the new spectrum research consortium. SEMATECH was an experiment in industry-government cooperation to strengthen the U.S. semiconductor industry. SEMATECH was created in 1987 by 14 U.S.-based semiconductor manufacturers and the U.S. government. The consortium allowed companies to leverage resources and share risks and costs in pre-competitive research. And it focused on improving industry infrastructure and capabilities and successfully reversed the serious decline in U.S. industry.[9]

Substantial private-sector involvement is crucial for the success of any new consortium of U.S. government agencies and private companies. Earlier collaborative efforts show that strong private-sector funding and direct involvement by senior industry management are crucial for success. While SEMATECH received government funding, private firms provided an equal share of the budget. Heads of companies served on SEMATECH's board. The same needs to be true for the spectrum research consortium. Private firms must be willing to participate and share resources for a research consortium to work. Well-designed U.S. support for research that supplements private-sector efforts and is done with minimal intervention in private-sector decisionmaking would enhance the overall U.S. research effort in spectrum technologies.

SEMATECH was an independent consortium. A new consortium could be organized independently, or it could be placed under the auspices of an organization like the National Science Foundation. In either case, the consortium will need to attract leading technologists and managers to a research center for periods of a few months to a year or more to provide a resource to the government, industry, and universities in the technical issues of the spectrum. It can act to generate new human capital for research through hiring and through arrangements with universities.[10]

The consortium could also provide a pool of talent to reinforce government expertise and provide independent studies on spectrum reallocation and changes in technology. Establishing a group of experienced, knowledgeable experts in an independent public institution could reduce shortcomings in research and technical expertise and better inform policy debates. This work should initially involve only U.S. scientists and engineers but should in the future expand to an international effort with the U.S. playing a leadership role.

A new research consortium for spectrum offers a cooperative approach to meeting the technical challenges of future spectrum-related research and the scope of resource requirements for it. However, implementing this recommendation

---

[7] Ben Charny, "Wi-Fi: Spelling Europe with an 'a,'" CNET News.com, June 28, 2002, http://rss.com.com/2100-1033-940352.html.

[8] Howell, "Competing Programs," pp. 240–241.

[9] International SEMATECH, "History of the Consortium," http://www.sematech.org/public/corporate/history/history.htm.

[10] Communication from William J. Spencer, chairman emeritus of International SEMATECH, April 2003.

poses significant challenges. It requires commitment of funds, recruitment of personnel, and development of an organizational structure before the consortium can take up its R&D-related tasks. This process will take both time and resources to achieve. The first steps might be to identify a federal sponsor for the effort, establish a small, "evangelical" steering group, and then organize private-sector participation in preparatory meetings. These are daunting tasks, but the alternative is to watch a continued decline in R&D, which will greatly weaken the United States. Public-private cooperation in research will be a key element for ensuring the strength and progress of U.S. industries, and our recommendation is that the United States use the SEMATECH model, suitably adjusted, to achieve this cooperation.

RECOMMENDATION 5

Create a National Spectrum Strategy

So our challenge is this: How do we fit new world-leading technologies into the U.S.'s own cramped spectrum allocation.—Secretary of Commerce Donald Evans

We join a number of studies on spectrum policy in calling for a national spectrum strategy.[1] National spectrum strategies are a part of the larger global restructuring of spectrum management. A strategy would replace the ad hoc approach used now and would provide a degree of predictability in place of the uncertainty that new technologies can create. Although the United States has a variety of spectrum strategies developed by different agencies and groups that are useful precedents, they do not provide a national vision for future spectrum use.[2]

A national strategy was unnecessary when there were fewer uses competing for spectrum and the technologies that used it were relatively homogenous. Increased demand, rapid technological change, and intense competition now make a national strategy essential, but these same factors will also complicate its development. Creating a national spectrum strategy will be difficult in an era of commercial uncertainty and technological change, and with the highly diverse and competitive communities that have an interest in spectrum matters.

The breadth and nature of these communities of spectrum users means that the development of a national strategy will need to be an inclusive advisory process. It will need to tap into the broad expertise available for spectrum. It will also need to actively involve the Congress. The task of creating the strategy should be given to the new White House spectrum management structure we have recommended, with the assistance of the senior advisory board, and for technological issues, a spectrum R&D consortium.

At the outset, we want to make clear that spectrum strategy cannot be a pseudonym for economic planning or industrial policy. We are not calling for a return to central planning or government-directed use of the spectrum. A national strategy that sought to impose a centrally planned approach or that attempts to

[1] A number of studies, including those of the GAO, Toffler Associates, and the Defense Science Board, cited previously, have called for the development of a national strategy for spectrum.

[2] Several agencies and organizations, including DOD, FCC, NTIA, and CTIA, have individually developed national strategies for spectrum.

closely control spectrum allocations would make matters worse, not better.[3] At the same time, an ad hoc or reactive approach to planning will no longer work. The United States cannot rely on market forces alone to achieve an optimal outcome for spectrum, and the new technologies that are reshaping spectrum use need both rules and a roadmap for implementation.

A national strategy will raise a series of issues related to new technologies and services. These include 3G communications, Wi-Fi, Ultrawide band, and digital broadcasting.[4] Beyond these technologies, there has been relatively little discussion of what a strategy should look like or do. Our view is that its primary function should be to map a course for transition and innovation and that a few broad issues will shape this:

- Balancing private and public-service spectrum needs and determining how to adequately protect safety-of-life services;
- Considering whether the national interest is best served by markets and expanded property rights, a commons model,[5] or government assignments of spectrum;
- Identifying areas and means for transition for both incumbent public and private spectrum users;
- Identifying spaces for experimentation and innovation.

Each of these raises a series of difficult subsidiary issues, including how to meet new demands while minimizing disruption to existing services; encouraging more efficient use of spectrum by both government and private-sector users; clarifying incumbent rights; meshing national priorities and international negotiations; finding ways to promote innovation and deciding where further deregulation is appropriate.

Balancing Private and Public-Service Spectrum Needs

Both the government and commercial sectors need better processes for reconciling critical governmental uses with consumer demands and commercial needs. Competition between public and private spectrum users is one of the driving factors for spectrum reform. Demand from both the public and private sectors outstrip the amount of available spectrum. This competition has been difficult for the United States to manage. Shifting spectrum between public and private uses will be one of the most politically charged tasks confronting a national spectrum strategy. The factors that will shape this task include assessing the effect of future demand and future technologies on spectrum use. It will include predictions of

---

[3] Martin Baily, Robert Willig, Peter Orszag, and Jonathan Orszag, *An Economic Analysis of Spectrum Allocation and Advanced Wireless Services* (Washington, D.C.: Sebago Associates, October 2001), provides an analysis of the problems of a centralized allocation process: information is not easily available, there can be significant political pressures and lobbying, and centralized processes are slow.

[4] Some analysts would say that 3G has crested and is being overtaken by new technologies. Whether this is true or not, the concern is that battles like that which occurred over 3G battle will become increasingly common in the future.

[5] The commons, a term drawn from medieval agricultural practices, refers to a resource owned by no one but available for use by all.

how international allocations could change. Finally, it will require estimates of the benefits and costs (including opportunity costs) of changing spectrum allocations.

The pressure created by these questions could be diminished by external factors. If the national strategy can identify flexible methods to transferring spectrum among private users, it will reduce pressure on the public sector to cede current holdings. Second, as innovation leads to new technologies that make more intensive use of spectrum, the pressure from increased demand will also be reduced. It will be important for a national strategy to find ways to move the United States out of the current environment of constraint and competition.

Protecting Safety-of-life Services

Identifying how to protect spectrum required to meet the nation's information needs and support safety-of-life functions will be a critical function for a spectrum strategy. The risks and costs of harmful interference rise drastically when safety issues are involved. If a routine cell phone call is interrupted because of interference, it is only inconvenient. Similar levels of interference to certain public safety, broadcast video and audio services, aviation and national defense functions could be catastrophic. An allowance must be made in any new spectrum-management paradigm for the protection of safety-of-life systems.

The traditional approach to protecting safety-of-life applications is for the government to assign frequencies protected by "guard bands"—zones of unused spectrum on either side. This reflects the limitations of past radio technologies and a management process that broke spectrum into an increasingly complex jumble of frequency assignments. For many sensitive government and information applications, this approach remains necessary. For other applications, guard bands are unnecessary and inefficient. A national strategy will need to identify both sets of applications and their requirements.

Markets or Commons

Debate over two very different regulatory models dominates public discussion of spectrum policy: whether to move toward a spectrum "commons" or toward a greater use of markets to assign spectrum. This debate will shape a national spectrum strategy, and the United States should draw on the experience of other countries to inform its efforts. Few countries have experimented with a spectrum commons, in part because the technologies that enable a commons are new and in part because of the complex regulatory changes required by a commons approach. In contrast, there is a long history of proposals to manage spectrum through competition and a number of these have been put into practice.

Australia reformed its spectrum management in the 1990s. The 1997 Australian Communications Authority Act merged the two agencies previously responsible for telecommunications and spectrum into a single unit. The new agency manages spectrum through planning and licensing and ensures compliance with licenses and standards. The act also created a class of license that, within assigned frequency bands, is "a tradeable, technology neutral (that is, the license is not related to any particular technology, system or service) spectrum access right for a fixed non-renewable term."[6] These licenses enable secondary markets for specified parts of the spectrum.[7]

This flexible licensing arrangement allows markets to reallocate spectrum to higher-value services.[8] This approach is appropriate to some, but not all, wireless services. The experience with mobile services provider Nextel shows both the benefits and risks of adopting flexible licensing like that used by Australia. Nextel (originally FleetCall) was able to buy specialized mobile radio (SMR) licenses in a series of markets and then use the spectrum to create a national cell phone network. Nextel shares the frequency band with state and local public-safety systems. Because the SMR licenses were originally allocated in 1974 for land-mobile services with a different architecture and pattern of use (mainly dispatch services), officials in several markets complained of interference from Nextel's mobile service, even though Nextel is operating in compliance with FCC rules.[9] A spectrum strategy would need to balance interference protection and licensing flexibility.

New Zealand restructured spectrum management in the late 1980s. The 1989 Radiocommunications Act created tradable long-term leases called "band management rights," which have no limit on the use of specific telecommunications or broadcasting applications. Certain frequencies are reserved for official use, but the effect of the act is to create a set of property rights. Private individuals can act as "band managers" who lease spectrum to others, but in practice, the New Zealand government has kept most management rights to itself and allocates management rights and spectrum licenses by way of public auction or tenders.[10]

The United Kingdom has gone furthest in using markets to change spectrum management, to the point of devising formulae that determine the amount that government agencies (including the Ministry of Defence) should pay for spectrum they use. The March 2002 *Review of Radio Spectrum Management*[11] concluded that

---

[6] See Australian Communications Authority, April 28, 2003, http://www.aca.gov.au/index.htm.

[7] The FCC recently approved a number of steps to authorize spectrum leasing and streamline the transfer of licenses. See "FCC Adopts Spectrum Leasing Rules and Streamlined Processing for License Transfer and Assignment Applications," May 15, 2003, http://hraunfoss.fcc.gov/edocs_public/attachmatch/DOC-234562A1.doc.

[8] See Hatfield, *Spectrum Issues for the 1990s*, chapter IV.

[9] Nextel Communications, "Promoting Public Safety Communications," white paper submitted to the FCC, November 21, 2001. Douglas Guarino, "Cell Signals Jam Cops' Radios," *Middleton (NY) Times Herald-Record*, June 17, 2002; "Interference Blocks Communications on County Transit Authority Radio," *Mobile Radio Technology Magazine*," June 4, 2002. Comments regarding the potential for interference were also provided as part of the FCC rule-making process by 25 cities and the states of Alabama, California, Florida, Hawaii, Maryland, and Michigan; Federal Communications Commission Docket 02-55.

[10] Ministry of Economic Development, "Radio Spectrum Auctions," http://www.med.govt.nz/rsm/auctions/index.html.

[11] Martin Cave, *Review of Radio Spectrum Management: An Independent Review for Department of Trade and Industry and HM Treasury* (London: Department of Trade and Industry, March 2002).

assigning spectrum through an administrative process "is no longer sufficiently flexible to meet the needs of the 21st century" and recommended:

- Selective deregulation of spectrum use;
- Increasing reliance on the market rather than administrative systems;
- Introduction of spectrum trading, combining new spectrum as it becomes available;
- A policy of reserving spectrum for public services such as defense, safety of flight or public safety, combined with a system of administrative charges to provide an incentive for government departments to economize spectrum use.

The UK says that these steps have increased the efficient use of spectrum and encouraged government agencies to shed unused or underused allocations. Although markets are a sore point for the telecommunications industry after the bursting of the information technology bubble and debt problems created by European 3G spectrum auctions, a national strategy will need to consider the benefits and risks of market mechanisms, including auctions, administratively assigned prices, and secondary markets. However, the United States will find it difficult to adopt policies that match the UK. The U.S. economy is larger; the Department of Defense makes more intensive use of spectrum, and telecommunications reform has progressed further in the United States, diversifying ownership and limiting government's ability to mandate drastic change.

No country relies entirely on market mechanisms for spectrum management, and market solutions do not work for all spectrum-management issues. The difficulties posed by the greater use of markets include how to estimate the price of nonprofit or public-service use of spectrum, potentially high transaction costs, and the need to coordinate national markets with international spectrum decisions. Markets are particularly inefficient in allocating spectrum between commercial and public uses like national defense. Some analysts also argue that the value of television and radio broadcasting is undervalued by markets.[12] Investors and incumbents also fear that markets would introduce significant new uncertainties into spectrum use.

A spectrum commons or "open spectrum" approach is the alternative to the greater use of markets and expansion of licensee rights.[13] FCC surveys of spectrum use conducted for the Spectrum Policy Task Force found that in many locations, spectrum use is highly sporadic, and much of the spectrum is not in use most of time. New radio technologies could exploit this unused spectrum or could operate below the "noise" threshold that already exists. Software-defined radio, for example, can automatically move among a wide range of frequencies to find and use temporarily unoccupied spectrum to transmit. Future advances in technology

[12] See Merton Peck, John McGowan, and Roger Noll, *Economic Aspects of Broadcast Regulation* (Washington, D.C.: Brookings, 1973), appendix A; and Bruce Owen and Steven Wildman, *Video Economics* (Cambridge, Mass.: Harvard University Press, 1992).

[13] Gerald Faulhaber and David Farber, two former FCC officials, have argued that expanded ownership rights, when combined with "easements" for certain uses, would allow the property rights and the commons approach to coexist.

could enable a spectrum commons where many devices use the same spectrum space without exclusive rights. Proponents of a commons model argue that instead of treating spectrum as a scarce physical resource, the United States should make many frequencies available to all as a commons.[14]

Assessing the benefits of the commons approach against both command and control processes and market mechanisms poses complex regulatory and political issues for national strategy. Advances in technology could make a spectrum-sharing approach an option for many, but not all, wireless applications. Accommodating a spectrum commons will require a reconceptualization of spectrum management. Skeptics also point to the need for regulatory structures to avoid a "tragedy of the commons" (i.e., the inefficient use of spectrum because of overcrowding and interference).[15]

Debate over a spectrum commons is also part of a larger discussion on how to encourage innovation. Innovation in information technologies underlies U.S. productivity increases in the last decade.[16] Continued innovation is crucial for the United States (which is why we have recommended the creation of a new spectrum R&D consortium), and the promotion of innovation must be a strategic goal. Many analysts and technologists argue that the commons model is the best way to promote innovation. Others contend that interference protection afforded by license rights is more effective. Mapping an approach that relies on experimentation and a progressive opening of spectrum to innovative uses, based on the result of that experimentation, could help reduce the uncertainty for investors and incumbents associated with the new technologies.

Since there is very little unencumbered spectrum, making space for innovation will affect the incumbents currently licensed to use spectrum.[17] Incumbents have made enormous expenditures to build the existing infrastructure in the United States and have deployed many innovative technologies. This infrastructure provides immense economic and safety benefits. It includes commercial activities such as cell phones and broadcasters and frequencies assigned to the military, the Federal Aviation Administration, and to local emergency services. A national strategy will have to balance the benefits of deploying new technologies that allow

---

[14] Kevin Werbach, "Open Spectrum: The New Wireless Paradigm," Spectrum Series Issue Brief #8, New America Foundation, October 2002. See also Michael Calabrese, "Battle over the Airwaves: Principles for Spectrum Policy Reform," Spectrum Series #1, New America Foundation, September 2001.

[15] Stuart Minor Benjamin, "Spectrum Abundance and the Choice Between Private and Public Control," Working Paper 03-3, AEI-Brookings Joint Center for Regulatory Studies, June 2003.

[16] The exact share of IT in the growth of productivity has been a matter of debate, but the contribution itself is no longer questioned. See Wessner, *Securing the Future*, pp. 21–24; and Stephen Oliner and Daniel Sichel, "The Resurgence of Growth in the Late 1990s: Is Information Technology the Story?" FEDS Papers, Federal Reserve Board, May 2000.

[17] The FCC's Spectrum Policy Task Force called for making more spectrum available for unlicensed use under Part 15 of the FCC's Rules. The FCC recently approved a Notice of Proposed rulemaking to do this. See "FCC Proposes Additional Spectrum for Unlicensed Use," FCC News Release, May 15, 2003, http://hraunfoss.fcc.gov/edocs_public/attachmatch/DOC-234566A1.doc.

greater use of the same spectrum by more users against the needs of incumbents and critical services.[18]

This balance will be shaped by the issue of spectrum property rights. Privately controlled resources have rights and protections that limit the government's ability (absent legislation or regulation) to affect them. Property rights and the rights of incumbents are central to the debate over how to change U.S. spectrum management. The 1934 Communications Act makes clear that licenses are for use, not ownership. Over time, however, the licensing process has created a presumption of ownership for many spectrum licensees. While the nation, not the licensee, "owns" spectrum, the idea of spectrum as a publicly owned resource has been eroded over the last decade.[19]

This accretion of quasi "property rights," where incumbents hold spectrum and have exclusive use but can trade or sell it only in limited circumstances, is one reason that the demand for spectrum outstrips the supply. The result is a set of precedents and expectations for incumbents that will complicate the development of a national strategy.[20] Deciding how to clarify property rights will be a major challenge for a national strategy. The strategy will need to balance the need to limit uncertainty, encourage aggregation of like uses, ensure consistency with international agreements, and promote innovation.

For the foreseeable future, the United States will need a strategy that blends government processes that assign spectrum, a greater use of markets for spectrum, and a commons model enabled by new technologies. The task for a national strategy will be to assess the trade-offs between these approaches. The national strategy will inevitably become an element in the competition over spectrum, but it will also provide a more transparent and orderly vehicle for resolving issues. The goal should not be a grand plan for spectrum use, but an iterative process with a regular cycle of review and revision that maps out approaches to better use of spectrum.

The recent FCC decision on ultrawide band (UWB) is a useful precedent for the national strategy.[21] Although concerns remain about the potential for interference from UWB devices, the decisionmaking process was open and saw good coordination, and ultimately, compromise among the many affected parties. It allowed different uses to share spectrum rather than creating new borders and assignments. A national strategy could build on the UWB precedent as a way to

---

[18] Japan, Norway, and Spain encourage innovation by assigning licenses for new services, sometimes for a relatively small fixed fee. Kenneth Neil Cukier and Justin Hibbard, "Spectrum Shortage: Giving away 3G Spectrum May Have More Merit than Auctions Do," *Red Herring*, October 2000.

[19] A Supreme Court ruling this year affirmed a lower court decision that licenses held by bankrupt wireless service provider NextWave were a company asset that the FCC could not reclaim. An 8 to 1 majority said that FCC authorities do not take precedence over bankruptcy claims. This confers the appearance of partial property rights for some license holders.

[20] Peter Cramton, Evan Kwerel, and John Williams, "Efficient Relocation of Spectrum Incumbents," *Journal of Law and Economics* (October 1998): 647–675.

[21] "The FCC's UWB Proceeding: An Examination of the Government's Spectrum Management Process," hearing before the Subcommittee on Telecommunications and the Internet of the Committee on Energy and Commerce, U.S. House of Representatives, June 5, 2002.

identify services that are important for national security or safety of life, such as the low-powered global positioning system signals transmitted from distant satellites, and to accommodate them in moving to the greater use of new technologies.

Conclusion

Spectrum management falls into a special class of political problems that results from technological change. To reap the full benefit of a new technology requires some level of reorganization, but reorganization is difficult. This was not the case in the nineteenth century, when government's regulatory role was minimal and private markets enabled reorganization as individuals chose to reallocate assets from old uses to new ones that they believed would yield greater returns. Government permission was seldom required.

This private approach led to two sets of objections. First, many argued that the market was wasteful and that central planning by experts was more efficient (this belief underlies many of the regulatory reforms of the first three decades of the twentieth century, including the establishment of the FCC). Second, the private approach often ignored costs to the larger society, especially costs to the public involving the environment, health, or safety. A long national debate led to two classes of regulation—one designed to allocate resources and the other designed to protect the public interest from harm.

Over the last two decades, the trend in regulatory practice under both Republican and Democratic administrations has been to reduce the role of government and rely more on private-sector decisionmaking. The U.S. regulatory structure has been reorienting itself away from the allocation of resources, while continuing to attempt to protect the public. Painful experimentation has proved that in most instances the market is ultimately more efficient than central planning for most resource or investment-related tasks. At the same time, well-designed regulations, by providing a structure for investment and a motivation for compliance, have proven effective in increasing overall efficiency and in reducing the risk of harm to the public from private activities. Continued regulation of spectrum by the FCC and NTIA will be necessary to provide this structure and ensure compliance.

Driven by the market and new technologies, political pressure for a new approach to spectrum management will continue to mount. Change in how the United States manages spectrum is both necessary and unavoidable, but reforming how the nation manages this valuable national asset will not be easy. Today's spectrum takes place in a complex regulatory and governmental environment that challenges the ability of an administration to undertake meaningful change. Our aim in making this set of self-reinforcing recommendations is to focus on pragmatic, achievable goals that will streamline the process for decision and reinforce consideration of broad national interests, so that the United States can begin now to take advantage of innovation and gain the full benefits of this immense economic resource.

A Checklist for Federal Spectrum Management Reform

1. Create NSC/NEC senior adviser.
2. Establish Spectrum PCC.
3. Create a National Spectrum Strategy.
4. Establish a Senior Advisory Board.
5. Create a public/private consortium for spectrum research.

Appendix A

Additional Views

David L. Donovan

At the outset, I want to compliment the members of the CSIS Commission on Spectrum Management and the CSIS staff for a thorough and reasoned report. The history of the government's approach to spectrum management illustrates the enormous difficulties in devising a rational spectrum policy for the United States.

There is little doubt that spectrum policy occupies an increasingly important role in our nation's economy. Spectrum use has also become important for our national security and the fight against terrorism. For example, Homeland Security Secretary Thomas Ridge stated that "obviously television and radio" are "the first choice" for disseminating information to the public during a terrorist attack.[1] The Office of Homeland Security recommends that a battery-operated radio or television be included in each home's emergency supplies.[2] The FCC established the Media Security and Reliability Council (MSRC), which has the task of insuring that broadcast facilities function properly in the event of a natural disaster or terrorist attack.[3]

The importance of over-the-air television broadcasting to this nation's information flow illustrates one of my fundamental concerns with this report. For broadcasting to function properly, there must be strict adherence to the interference rules that underpin the system. There are more than 280 million television sets in the hands of consumers today, and more than 25 million new sets are sold annually. Approximately 81 million of these television sets rely exclusively on over-the-air reception. With the enactment of the FCC's digital television (DTV) tuner rules, the sales of new DTV television sets with off-air tuners will accelerate rapidly. Accordingly, interference concerns will be especially acute during the next few years as the entire broadcast system completes the transition to digital television. To make matters more complex, we are just beginning to gain "real world" experience with the interference characteristics of digital

[1] PBS Online News Hour, Newsmaker: Tom Ridge, February 19, 2003, at http://www.pbs.org/newshour/bb/terrorism/jan-june03/ridge_2-19.htm. "JIM LEHRER: [S]ome people have mentioned that how is the ordinary American to find out about a terrorist attack…? Is there some kind of system being worked on for that? TOM RIDGE: Precisely. There are multiple ways that we can communicate the plan; but there are also multiple sets of circumstances under which some of them wouldn't work. And so obviously television and radio is our first choice…. [I]f the electricity is off, hopefully a battery-powered radio might help."

[2] See U.S. Department of Homeland Security, "Make a Kit," at http://www.ready.gov/supply_checklists.html (last visited April 17, 2003).

[3] Bill McConnell, "Ridge Takes the Point," *Broadcasting and Cable*, June 2, 2003, p. 28.

broadcasting. Great care must be taken to insure that the American public continues to enjoy this service in today's dynamic video environment.

The report recognizes that incumbents, such as broadcasters, are concerned with potential interference from sharing spectrum with new wireless devices. It suggests that new regulations may be needed to require manufacturers to build better receivers and transmitters. Nonetheless, the report also suggests that we measure interference in new ways and define what levels are "acceptable." Although it is not entirely clear, the report seems to suggest that incumbent users may have to accept new levels of interference in order to accommodate the demand for new licensed and unlicensed wireless devices. American consumers have already spent billions of dollars on existing equipment and are in the process of acquiring new digital receivers. They are unlikely to accept the notion that their television sets may malfunction because the government redefines "acceptable" or "harmful" levels of interference. The problem associated with existing legacy equipment cannot be swept aside by "new" definitions.

Moreover, basing spectrum policy primarily on rules designed to control the interference-immunity characteristics of wireless devices appears to be a risky proposition. This is especially true with respect to unlicensed wireless devices. If such devices ultimately cause interference, they may render unusable existing spectrum-based services and equipment. Once distributed to the general public, there is no practical way to locate or reclaim defective unlicensed devices. Such "defective" devices, whether due to improper design or manufacturing error, not only cause interference to existing services but also may limit the future uses and reduce spectrum values.

One of this nation's highest spectrum priorities should be to ensure that the transition to digital broadcasting occurs in a timely and economic manner. Completing the digital transition will "free-up" 108 MHz of spectrum for other uses, and will help address the increasing demand for spectrum referenced in this report. One key method of insuring a timely transition is to preserve the spectrum integrity of the licensed television broadcast service. This means that new, off-air digital television receivers must remain free from interference that may be caused by new wireless devices. From a spectrum efficiency perspective, increasing interference levels in the broadcast band during the digital transition is counterproductive. Rather than promote spectrum efficiency, it will merely delay the final transition and the return of the 108 MHz of spectrum.

I focus on this issue because accommodating the needs of new wireless devices appears to be a primary justification for the structural recommendations contained in this report. However, there appears to be no necessary nexus between accommodating new wireless devices and the need for structural reform. For example, the need for a uniform national spectrum policy, especially as it relates to this nation's international negotiations is, by itself, a sufficient justification for some of the recommendations made in the report. Alternatively, under the current spectrum policy regime, the FCC's Spectrum Policy Task Force has taken the lead in fostering the development of new wireless services and devices. Finally, although the process is time consuming, licensing spectrum gives the government the flexibility to alter spectrum uses to meet new demands. Such flexibility may become limited in an environment with millions of unlicensed devices operating on certain frequencies.

The report's recommendation that spectrum policy occupy a more prominent role within the executive branch is correct. President Bush's executive memorandum revising the federal government's approach to spectrum management policy is an important step in policy reform. Moreover, earlier this year the FCC executed a new memorandum of understanding on spectrum coordination with the Commerce Department's National Telecommunications and Information Administration.

We must be careful, however, not to usurp congressional prerogatives in this area. Greater coordination and spectrum policy leadership in the executive branch is essential for a national spectrum policy. Nonetheless, this report's recommendation that the White House should seek legislation to subordinate all aspects of spectrum management in the executive branch, if cooperation between the FCC and the new White House spectrum policy structure becomes "unwieldy," may go too far. Given the history of close cooperation between the FCC and the executive branch, we need not reach this conclusion at this time. Moreover, unlike military base closures, which ultimately concern the need for an efficient nationwide defense capability, spectrum issues have a significant local component. Local public safety officials need spectrum to address local emergencies. Broadcast stations serve as an outlet for local public expression, and they are critical to local political discussion. Accordingly, locally elected members of Congress have a legitimate role to play in this nation's spectrum policy.

On balance, the commission's report represents a serious effort toward resolving some critical issues affecting the nation's spectrum policy. Many of its recommendations will help create a process to resolve the complex spectrum issues that now confront government and business leaders. This report will help shape the spectrum policy debate in the years ahead.

Raul R. Rodriguez

I would like to thank CSIS for offering me the opportunity to participate in the work of this commission, comprising so many talented and accomplished individuals. And, I wish to thank our able cochairmen for their guidance and their vision and for the enthusiasm they brought to the work of the commission. It was my pleasure and honor to have worked with each of you.

The commission's report identifies several aspects of U.S. spectrum management that need to be improved and provides viable options for policymakers to consider. In general terms, I concur in most of these recommendations. They should help stimulate thinking and dialogue and, hopefully, lead to positive changes in the manner in which we manage and allocate spectrum in the United States. I must, however, respectfully demur on two of the report's recommendations.

I cannot align myself with a recommendation that calls for the creation of a joint government–private-sector group whose purpose would be to undertake research and development of commercial wireless technologies. No case has been presented that there is a need for departure from our current practices. Research and development of commercial products ought to remain a private-sector initiative without government interference or largesse. If government policymakers believe they need to stimulate research in a given field, there are many means of achieving that end, including enacting tax incentives to stimulate private investment in the desired field of study. If the military believes it needs to develop a particular technology, it has the means and the experience to accomplish that or to pay to have it done. In my estimation, what is suggested in recommendation 4 is counterintuitive to our national economic model; it is a classic example of industrial policy setting, which that would result in a government-sanctioned method of choosing technology winners and losers. This is something best left to the markets to decide.

I also cannot concur in a recommendation to create a permanent WRC ambassador position at the Department of State. I have known personally and worked well with every head of CIP since it was first created two decades ago. Each has served exceptionally as our industry's ambassador-at-large and as the head of occasional telecom conferences. However, it would be very disruptive of his/her duties and responsibilities for the head of CIP to undertake months of planning and preparation for major ITU conferences and then to spend one month or longer away from his/her office to be present at an ITU conference.

The problems we have experienced in the past with the WRC ambassador are twofold and can be remedied without having to change the role of the head of CIP. The two problems have been: (1) White House failure to name the ambassador in a timely manner; and (2) inadequate funding to carry out the responsibilities of the office. Both of these problems can be remedied without naming a permanent ambassador. I am not convinced the problems would not remain even if the position were made permanent. We should not lose sight of the fact that the United States historically has achieved its goals at WRCs, much to the credit of the one-time-only heads of delegation that have owed allegiance to the president who appointed them and not to bureaucrats at any one agency.

Appendix B

Participants List

Cochairs

Robert Galvin
Former Chairman, Motorola Inc.

James Schlesinger
Former U.S. Secretary of Defense

Project Director

James A. Lewis
Director, Technology & Public Policy
CSIS

Members

Steven Berry
Senior Vice President
Cellular Telecommunications & Internet Association

David Beier
Partner
Hogan and Hartson

D. James Chadwick
Director of Spectrum Management
MITRE

Brian Dailey
Corporate Vice President
Lockheed Martin Corporation

David L. Donovan
President
Maximum Service Television, Inc.

Dean J. Douglas
General Manager, Mobile e-business Services
IBM

John Douglass
President
Aerospace Industries Association

Craig Fields
Chairman
Defense Science Board

John Hamre
President & CEO
CSIS

William G. Howard Jr.
Chair
Defense Science Board Task Force on Spectrum

Paul Kaminski
Former Chairman
Defense Science Board

Travis Marshall
Former U.S. Ambassador to the World Radio Conference

Dave McCurdy
President
Electronic Industries Alliance

Gary Minden
Professor
University of Kansas

Carl O'Berry
Senior Vice President
The Boeing Company

Janice Obuchowski
President
Freedom Technologies, Inc.

Clyde Prestowitz
Founder & President
Economic Strategy Institute

Raul R. Rodriguez
Partner
Leventhal, Senter & Lerman

Gail Schoettler
Former U.S. Ambassador to the World Radio Conference

Thomas Von Essen
Former Commissioner
New York City Fire Department

Richard E. Wiley
Partner
Wiley, Rein & Fielding

Experts

Michelle Farquhar
Hogan & Hartson
Former Head, FCC Wireless Bureau

Cathy L. Slesinger
Senior Vice President
Cable & Wireless

Gerald Musarra
Vice President
Lockheed Martin

Jennifer Warren
Vice President
Lockheed Martin

Bruce Mahone
Director, Space Policy
Aerospace Industries Association

Steve Sharkey
Director, Spectrum & Standards Strategy
Motorola

Observers

Steven Price
Former DASD-Spectrum/C3 Policy
Office of the Assistant Secretary of Defense

David Gross
U.S. Coordinator, International Communications and Information Policy

Michael Gallegher
Deputy Assistant Secretary for Administration
NTIA, Department of Commerce

Paul Kolodzy
Former Cochair
FCC Spectrum Task Force

Peter Tenhula
Senior Legal Adviser
FCC

Gil Klinger
Director, Defense Policy & Arms Control
National Security Council